THE YOUNG
AND HIS TELESCOPE

THE
YOUNG ASTRONOMER
AND
HIS TELESCOPE

PATRICK MOORE

KEITH REID LTD
SHALDON   DEVON

ISBN Cased edition 0 904094 08 1

ISBN Limp edition 0 904094 09 X

Printed in Great Britain by
Devonshire Press Ltd, Torquay

Acknowledgements

I am most grateful to David Lowne, for acting as 'demonstrator' and who is shown in all the photographs in this book. The pictures were taken in March 1974, at Selsey.

My thanks also to Lawrence Clarke for his drawings, and to Keith Reid for his invaluable help throughout.

PATRICK MOORE
Selsey   May 1974

Contents

To
KATIE

List of Illustrations

1 Beginning

When I first became interested in the stars, at the age of six, astronomy was thought to be a rather unusual hobby for a boy. Even when I began my 'Sky at Night' television series for the British Broadcasting Corporation, twenty-eight years later—that is to say, in 1957—not many people took a really serious interest. Now, in the Space Age, everything has changed. We have become used to the idea of men on the Moon, probes to the planets, and artificial satellites circling the world; we use space research methods to beam television programmes direct from one side of the Earth to another, and there is serious talk of setting up a full-scale lunar base before the end of the century.

Astronomy has come 'down to earth' with a vengeance, and almost everyone takes at least some notice of what is happening in the night sky. Inevitably, I receive a great many letters asking for my comments and advice; and this book is a deliberate attempt to answer some of them. Questions from enthusiasts of school age (that is to say, from about six up to about seventeen) tend to repeat themselves over and over again, and the commonest of all are as follows:

1 How do I start taking a proper interest in astronomy?
2 Where can I buy a good, cheap telescope?
3 How do I set about a school project on astronomy?
4 Can I make a career out of astronomy—and is it necessary to take a science degree?

I have long since lost count of the number of inquiries I have had about these four points. Therefore, perhaps it is time that I tried to sum matters up as neatly as I can. Let me make it clear, at the outset, that this is not meant to be a general introduction to astronomy. All that I propose to do is to answer these four quite definite questions.

Very well, then: what is the first step?

Most people know that the Earth is a planet—a world almost 8000 miles in diameter, moving around the Sun. The Earth-Sun distance is, on average, about 93 million miles, which is a vast distance by everyday standards but does not seem much to an astronomer. The Sun itself is a star, no brighter or larger than many of those we can see on any clear night; it seems so splendid in our skies merely because all the other stars are so much further away. The Sun is the ruler of what we call the Solar System, which includes eight planets as well as the Earth, together with various bodies of lesser astronomical importance—such as the Moon, which is our companion in space, and stays together with us as we travel round the Sun. The brightest planets are Venus, Mars, Jupiter and Saturn. Like the Moon, they have no light of their own, and shine only because they reflect the rays of the Sun.

This may be the limit of knowledge for the absolute beginner, and before he can go any further he must clearly learn a little more. What I did, while still a small boy, was to read books. The immediate difficulty here is that books are expensive. Also, there are so many books published that it is not easy to make a choice. On page 76 I have listed some that ought to be of help, but please bear in mind that the list is very incomplete, and is bound to reflect my personal views— with which not everyone will agree. Also, new books come out frequently, so that the list will soon become out of date.

To buy even a selection of books would cost many pounds, which is certainly more than most school-age readers can afford. Luckily, there is a solution. Every town has its public library; and in most cases there will be at least half a dozen books there to be read, or borrowed for more careful study. If you want a definite title, and it is not to be found, the library will probably be able to obtain it; this is what libraries are for. So for a start: Go to your local library, see what is available, and do some reading.

At the same time, start taking a practical interest. You do not need a telescope, or indeed anything except your eyes; the only real extra requirement is either a simple star-map, or else somebody who knows the star-patterns and can point them out. Because the stars are so far away they seem to stay in the same relative positions for year after year, century after century (at least from the point of view of the naked-eye observer), and their patterns or constellations are much easier to identify than might be thought. For instance, it is by no means difficult to pick out the seven fairly bright stars which make up the Great Bear, otherwise nicknamed the Plough, or (in America) the Big Dipper. From Britain, the Bear is always visible whenever the sky is clear and dark; and once you have found it, you will always be able to identify it again. Sometimes it will be overhead, sometimes rather low down; but its shape will never change.

My own method of learning my way around the sky was to find a few of the easiest constellations, and then use them to show the way to the others. For instance, two of the Bear stars point to the Pole Star, which always lies due north (or virtually so) and from Britain is fairly high up. The

13

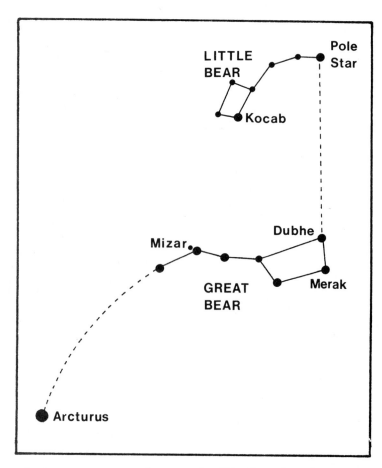

THE GREAT BEAR AS A 'SIGNPOST'. *The Great Bear (or Plough) is always visible from Britain whenever the sky is clear and dark. It can be used as a guide to other stars and constellations; for instance, the Little Bear, which contains the Pole Star, and also the brilliant orange Arcturus, in Boötes, which however does not remain above the British horizon all the time.*

14

diagram given here shows what is meant. When the Pole Star has been found, it is easy to track the much dimmer outline of the Little Bear, which curves down over the Great Bear's tail. Next, follow round the tail itself until you come to a very brilliant orange star; this is Arcturus, in the constellation of Boötes, the Herdsman. Unlike the Bear, it is not always above the British horizon; but when it is, you will have no trouble in locating it.

There are many other groups which can be found from the Bear alone, and in winter we also have the magnificent constellation of Orion, the Hunter, which has a very distinctive shape; from late autumn until early spring it dominates the southern part of the night sky. Of its two brightest stars, one (Betelgeux) is orange-red, while the other (Rigel) is glittering white. In the centre part of Orion there are three stars lined up, forming the Hunter's Belt; downwards they show the way to Sirius, the brightest star in the sky, while upwards they point to another red star, Aldebaran, in the constellation of the Bull. Follow the Belt-to-Aldebaran line still further, curving it somewhat, and you will come to the hazy patch which marks the cluster of the Pleiades or Seven Sisters; and so on.

I could give many more examples; but, as I have said, I am not trying to write a book about general astronomy, and it is enough to suggest the method. Try it out, and you will find that it works. If you make a resolve to identify one new constellation every clear night, you will very soon find that the star-patterns become really familiar; and the stars themselves become far more interesting when you know which is which.

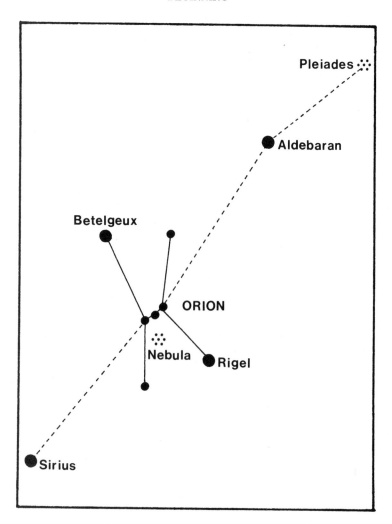

ORION. *Orion dominates the southern part of the sky during winter, and it also can be used as a guide; shown here are Sirius, Aldebaran, and the star-cluster of the Pleiades, or Seven Sisters.*

16

To sum up:

1 Read some books. If you do not have any, and do not want to go to the expense of buying them, consult your local public library.

2 Obtain a star-map, which is cheap and easily found (see page 77). Spend some time outdoors on a clear evening, and start learning your way around the constellations.

And then—if you are still interested—consider what can be done about equipment.

2 The Astronomer's Telescope

Before coming on to the all-important question of choosing a telescope, I must say something about the way in which telescopes work. There are two main types: refractors and reflectors. Each type has its own advantages, and its own drawbacks.

With the refractor, the light from the object to be studied is collected by a specially-shaped glass lens, known as an object-glass or objective. The light-rays are bunched up, as shown in the diagram, and brought to a focus. The image is then magnified by a second lens, called an eyepiece. Note that it is the eyepiece which is responsible for the actual

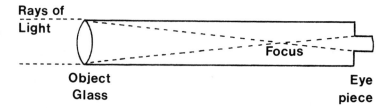

Rays of Light

Object Glass

Focus

Eye piece

PRINCIPLE OF THE REFRACTOR. *The light is collected by an object-glass and is brought to focus; the image is magnified by an eyepiece. The distance between the object-glass and the focus is called the focal length, and this, divided by the diameter of the object-glass, is the focal ratio. Thus, if the focal length is 36 inches and the object-glass is 3 inches across, the focal ratio is 36/3=12, written simply as f/12.*

magnification; all that the object-glass does is to collect the light—and, of course, the larger the object-glass, the more light it can gather in. Thus a 3-inch refractor (that is to say, a refractor with an object-glass 3 inches in diameter) is much more effective than a 2-inch. By changing the eyepiece, you can alter the magnification.

All astronomical refractors give an inverted or upside-down image. With a telescope of the kind used for looking at ships out to sea, an extra lens-system is included to turn the image the right way up again; but this means losing a certain amount of light—which is what the astronomer is particularly keen to avoid; after all, it does not in the least matter whether the Moon or a planet is seen upside-down or not. Correcting lenses are, however, always used in binoculars. A pair of binoculars is nothing more nor less than two small refractors used together, and I will have

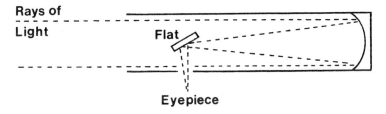

Rays of Light

Flat

Eyepiece

PRINCIPLE OF THE REFLECTOR. *The light is collected by a mirror, and is directed back up the tube on to a smaller flat mirror; the focus is formed toward the side of the tube, and the image is magnified by an eyepiece. The focal ratio is, as before, the distance between mirror and focus; most small reflectors are of the f/6 to f/8 variety, though some are longer (my 12½-inch reflector is f/6.) This is the Newtonian pattern. There are others, but the Newtonian is the commonest among amateurs.*

19

much more to say about them later.

The reflecting type of telescope has no lens. Instead, the light passes down an open tube and falls on to a curved mirror; the rays are then sent back up the tube on to a smaller 'flat' mirror, placed at an angle. The rays are sent into the side of the tube, where an image is formed and magnified by the eyepiece. With a reflector of this kind, known as a Newtonian (because the principle was first used by Sir Isaac Newton, more than three hundred years ago), the observer looks into the tube instead of up it. The effectiveness of the telescope depends on the size and quality of the mirror.

Now for the question: 'Where do I buy a cheap telescope that is of real use in astronomy?' And the answer—which I give with regret—is quite straightforward: 'You don't'. The honest truth is that a useful telescope is an expensive item if bought new. It will cost at least £50 and probably more.

There are plenty of cheaper telescopes available; and here I am going to be deliberately controversial. In my opinion, the minimum useful size is 3 inches aperture for a refractor, and 6 inches for a reflector. (Inch for inch, an object-glass is more effective than a mirror.) It is quite true that a smaller instrument will give very pleasant low-power views of star-fields and the craters of the Moon; but it will probably have rather a small field, and it will not be powerful enough to give real satisfaction for long. Though many people will disagree with me, I would not personally be happy about paying a relatively high price for, say, a 2-inch refractor or a 4-inch reflector.

Until fairly recently it was possible to buy good second-hand telescopes; and it is worth remembering that if a telescope is well looked after, it will not deteriorate with age.

(I have an excellent 5-inch refractor which must have been made about 1880, and is just as good now as it was then.) Unfortunately this is no longer the case, and second-hand telescopes have become decidedly rare. Moreover, it is necessary to take great care of them. If the object-glass of a refractor, or the mirror of a reflector, is of poor quality the instrument will be useless; and a bad telescope is not always recognizable on sight. Unless you have a real knowledge of optics, never buy a second-hand telescope before trying it out or having it checked by an expert.

Of course, bargains can be found sometimes. Only a few months ago a schoolboy called to see me bringing with him a well-mounted 2-inch refractor which he had been offered for the modest sum of £10. I tried it out, and found that within its limitations it worked very well; it gave excellent small-scale views of the Moon, and it also performed well when I turned it toward star-fields. At £10 it was obviously worth buying, and it has given its owner a great deal of pleasure. But this is exceptional; and while I gladly recommended it at £10, I would have been much more hesitant if the price had been, say, £30. The same applies to reflectors with mirrors below 6 inches in aperture.

There is another point to be borne in mind, too. There are some people who have no wish to do more than look casually at the night skies; they are content to have a telescope which will give pleasant, low-magnification views of the Moon and star-fields. If this is the case, then buy a very small telescope by all means. But if you want to do some real observing, something larger is absolutely essential.

All this may sound depressing, and I can well imagine our would-be astronomer saying: 'If it means spending something between fifty and a hundred pounds, I might as well give up at once.' Fortunately there is no need for this. If you have only a limited amount of money (as most people have nowadays!) the solution is to postpone buying a telescope, and make do, as a start, with binoculars. Here the financial outlay is much less; and a pair of good binoculars has most of the advantages of a very small telescope, apart from sheer magnification.

My suggestions, then, are as follows:

If you are no more than casually interested, a 2-inch refractor will serve you quite well. If you want to take matters further, then begin by buying binoculars. They will give you tremendous pleasure; and in the meantime you can consider whether you do, in fact, want to save up for a telescope powerful enough for you to undertake some real astronomical observing.

3 Binoculars in Astronomy

As I have said, a pair of binoculars is made up of two small refractors joined together, so that both eyes can be used. There is a single focusing arrangement, though in most binoculars it is also possible to focus one of the 'telescopes' independently so as to give a really sharp, clear-cut image. The object is seen right-way-up, which means that binoculars can be used both for astronomy and for ordinary, everyday use.

The price is very much lower than for even a small astronomical telescope. If you have £15, you can obtain a really good pair, and it is often possible to buy satisfactory binoculars for even less. This time there is no objection to a second-hand purchase, because the binoculars can be tested at once, and any really bad faults will be all to obvious. Railway 'lost property' shops can be useful here; the number of binoculars left in trains, and never claimed, is quite remarkable.

Binoculars are classified according to their magnification and the diameter of each object-glass. Thus a 7 × 50 pair has a magnification of 7 times, with each object-glass 50 millimetres in diameter. Once again, the larger the aperture, the more light can be collected; but there are several important points to be borne in mind. Higher magnification means a smaller field of view, and less ease of use.

Generally speaking, I would say that the 7 × 50 type is more or less ideal. It will have a wide field, and will gather in a satisfactory amount of light; the binoculars will not be too heavy to hold steady—and when looking at an astronomical body, steadiness is all-important. If the object is wobbling about (or, to be more accurate, if your binoculars are

USING BINOCULARS. *The binoculars which David Lowne is using in this photograph are of the 7 × 50 variety; these allow for a wide field of view, together with sufficient light-grasp, and they can be held comfortably. They are excellent for viewing star-fields, and will show the craters and mountains of the Moon, the phases of Venus and the four principal satellites of Jupiter. (Remember, however, never to turn them toward the Sun.) They can also be used for daytime viewing—ships out at sea, for instance; bird-watchers also favour them. With increased magnification the problem of steadiness becomes more and more evident. With a magnification of over about 12, some sort of mounting is desirable, and for a magnification of 20 it is a necessity.*

wobbling about) it will be impossible to see any detail at all, and the unsteadiness problem grows rapidly with increased magnification.

I have several pairs of binoculars, and have tested them out very thoroughly. 7×50 is excellent both for the sky and for terrestrial use, such as looking at distant ships or for bird-watching. I have had no difficulty in holding my 8×50 binoculars steady, and not very much with my 10×60; but with a magnification of 12 the trouble starts to become evident. It is also possible to buy much more powerful binoculars—say 20×70, but the instrument is heavy, and to hold it still enough to examine a star-field or the Moon is very difficult. Anyone who wants a magnification of this kind will have to make a mounting for the binoculars; but I do not propose to say more here, because 20×70 binoculars cost much more money, and we are trying hard to keep the cost down.

In short: I would recommend a magnification of between 7 and 10, though 12 is tolerable—in which case the aperture of each objective will be somewhere between 40 and 60 millimetres. You will be very lucky to find a good pair at less than £12, even second-hand, but it should certainly be possible to buy excellent binoculars for something between this and £15. The thing to check, before completing a purchase, is sharpness of focus. If the image is blurred in either eye, then look elsewhere.

Armed with 7×50 binoculars (or something similar), the observer will find that his scope is broadened beyond all recognition. But one thing he must *not* do, under any circumstances, is to look at the Sun—or anywhere near it. The reason should be obvious enough. Not only the light is focused, but also the heat; and the result of concentrating the Sun's heat on to your eyes, even for a fraction of a second, is certain and permanent blindness. I have no idea how many times I have given this warning, both in writing and on radio and television; but I make no apology for repeating it yet again. Even when the Sun is low over the horizon, shining through mist or thin cloud and looking deceptively harmless, keep your binoculars well away from it.

The Moon is a different matter, because there is to all intents and purposes no heat; and even if you dazzle yourself (as may happen near full moon) no harm is done. And through binoculars the Moon is indeed a lovely sight. It is a world roughly a quarter of a million miles away from us, and much smaller than the Earth; its diameter is a mere 2160 miles, and its weak pull of gravity means that it has no atmosphere, so that its surface features are sharp and clear-cut. Its phases, or apparent changes of shape from new to full, are due to the fact that only half of it can be sunlit at any one time; and only at full moon is the whole of the sunlit hemisphere turned toward us. At other times the Moon appears as a crescent, half, or three-quarter ('gibbous').

Oddly enough, full moon is the very worst time to start observing, because the sunlight on it is coming 'straight down', so to speak, and the mountains and crater-walls cast almost no shadows; all that can be made out are the particu-larly dark areas. The vast grey plains, still called 'seas' even though there is no water in them, are obvious enough with the naked eye, but binoculars show them much more distinctly, and near full you can also see the bright streaks or rays centred upon a few of the brightest craters. The view when the Moon is away from full is much more spectacular. Look at the 'terminator'—that is to say, the boundary between the sunlit and the night hemispheres—and you will see the peaks and the craters magnificently. It is fascinating to check the view from one night to the next; the changes in the angle at which the sunlight hits the Moon cause striking alterations.

Another thing worthy of note is the earthshine. When the Moon is at its crescent stage, the 'night' side can often be seen shining dimly—because it is being illuminated by light reflected on to it by the Earth. With binoculars, considerable detail can often be seen in the earthlit portion, and there is one particularly brilliant crater, known as Aristarchus, which generally shows up as a bright patch. An outline map of the Moon will help in identifying the main surface features on the sunlit hemisphere; I will have more to say about this later.

When the Moon is near full it tends to drown the stars and the planets; so let us now assume that the night is dark and moonless, so that we can use our binoculars to best advantage. What are the first things to study?

Planets, in general, do require higher magnification than ordinary binoculars can provide; but they show up as disks rather than as points of light, so that they are quite unstarlike. The brightest of them is Venus, which is about the same size as the Earth, and is closer to the Sun than we are (67 million miles, as against our 93 million). Because it depends entirely upon reflected sunlight, and does not always keep its 'day' side turned toward the Earth, it shows phases similar to those of the Moon; and when Venus is a crescent, the form is very easily visible in any pair of binoculars with a magnification of 7 or more. When Venus is visible in the evening sky after dark, in the west, its phase will decrease steadily—that is to say, it will change from a

27

half into a crescent, narrowing until the planet has become too near the Sun to be seen at all. (I repeat: do not sweep around for Venus until after sunset.) At other times Venus appears in the eastern sky before dawn; this time the phase increases, so that the crescent stage is followed by a half-moon appearance and then a three-quarter shape. When Venus is full, with the whole of its illuminated hemisphere facing us, it is of course on the far side of the Sun, and to all intents and purposes out of view.

Mars, the Red Planet, shows no detail in binoculars other than its tiny disk; but Jupiter is more rewarding. Here we have the largest planet in the Solar System—a huge globe more than 88,000 miles in diameter. It is attended by a whole family of moons or satellites, of which the principal four are both large and reasonably bright. 7 × 50 binoculars will show them; they appear as 'stars' close beside Jupiter itself, and their positions relative to the planet change from night to night. I have never had much trouble in seeing them with my 7 × 50 binoculars, though they are admittedly much more obvious with a higher magnification.

Saturn, second of the giant planets, is surrounded by a superb system of rings, but unfortunately these cannot be properly seen with binoculars, and the best that can be done is to make out that there is something unusual about Saturn's shape. So let us pass on to the true stars, where the scope is endless.

Any careful glance at the sky will show that the stars are not all alike. No telescope yet built will show a star as anything but a dot of light—they are too far away—but they are quite clearly of different colours. I have already referred to the two leaders of Orion, Betelgeux and Rigel, which are as unlike as they could possibly be. The orange-red hue of

28

Betelgeux is due to the fact that its surface is much cooler than that of the white Rigel; after all, white heat is hotter than red heat! (To make up for this Betelgeux is a giant by any standards; its diameter is 250 million miles, so that it could swallow up the whole path of the Earth round the Sun.) Another red star is Aldebaran, in Taurus, which is more or less lined up with Orion's belt; and we also have Antares, in the southern constellation of the Scorpion, which is well seen above the horizon during summer evenings. On the other hand Vega, which lies almost overhead during evenings in summer, is distinctly blue.

With the naked eye, only the very brightest stars show obvious colours, but binoculars show that the fainter stars too have their own particular hues. Look, for instance, at the two Pointers in the Great Bear. The fainter one, Merak, is white; but the other, Dubhe, is quite definitely orange when observed through binoculars. Another orange star is Kocab, in the Little Bear, also shown in the diagram on page 14. These are only a few examples; range round with your binoculars, and you will find many more stars which are obviously not white.

Incidentally, it is worth noting that star-twinkling is due entirely to the Earth's atmosphere, which is unsteady and which affects the starlight passing through it. A star which is low over the horizon will twinkle much more strongly than a star which is higher up, because its light is coming to us through a thicker layer of air. Look, for instance, at Sirius, which is the most brilliant star in the whole sky, and is always rather low down as seen from Britain. It is really pure white; but it seems to twinkle violently, showing various colours and looking like a flashing jewel. Binoculars give a beautiful view of it.

Then, too, we have double stars—genuine pairs, made up
of two stars which are associated with each other. Mizar, in
the Great Bear, is probably the most famous of them. When
the sky is clear and dark, a fainter star, Alcor, may be seen
close to Mizar; binoculars show it much more clearly—and
with a telescope it is found that Mizar itself is made up of
two, so close together that to the naked eye (or in low-power
binoculars) they appear as one.

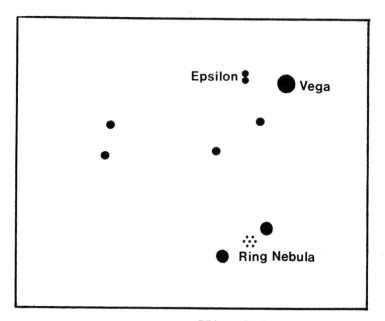

POSITION OF EPSILON LYRÆ.   *This multiple star lies near the
brilliant Vega. Binoculars will show that it is double; a small
telescope—say a 6-inch reflector—will show that each component is
again double. The Ring Nebula, also near Vega, is rather faint
with small telescopes.*

An even better example for the binocular-owner is Epsilon Lyræ, close to the brilliant blue Vega. Here we have a case of stellar twinning; the components seem equal in all respects. Keen-sighted people can split Epsilon Lyræ with the naked eye, but I think that most observers will need binoculars to see them really clearly.

Again these are only two examples out of many, and the number of double stars within binocular range is surprisingly great. We also have variable stars, which—as their name suggests—do not shine steadily; they brighten and fade, so that their alterations can sometimes be noticed even from one night to the next. But variable star observation brings us to more serious astronomy, and I will say more about it when I discuss school projects.

It is with star-fields and star-clusters that binoculars really come into their own. There is, for instance, the Milky Way, that glorious band of light which stretches across the sky from one horizon to another. City-dwellers will not see it; the glare of street-lamps and illuminated advertisements will drown it completely, but in the darkness of the country-side it is a glorious sight. Turn binoculars toward it, and you will see that it is made up of stars—so many, and so apparently close together, that to count them would be quite impossible.

In fact, the stars in the Milky Way are in no danger of bumping into each other. Our star-system, or Galaxy, is flattened in shape, with a central bulge; and when we look along the main thickness of the Galaxy we see many stars in almost the same line of sight, producing the Milky Way aspect. Altogether the Galaxy contains something like a hundred thousand million stars, each of which is a sun in its own right. Our Sun lies well away from the centre, but not very far from the main plane.

31

It is well worth sweeping along the Milky Way with binoculars; but there are also separate star-clusters, of which the most famous is known as the Pleiades or Seven Sisters. There is no problem in finding it, because it is very conspicuous even with the naked eye; it is shown in the diagram on page 16, in the constellation of Taurus (the Bull), and is on view from early autumn through to the spring. At least seven separate stars can be seen in it without optical aid under good conditions, but binoculars will show many more. In the Pleiades the brightest members of the cluster are hot and white, and much more luminous than the Sun.

Because the Pleiades cover a fairly wide area, telescopes will show only a part of the cluster at any one time, and the main beauty is lost. There is no doubt in my mind that binoculars give the best possible view of them, and this also applies to the much more scattered cluster of the Hyades, which lies round Aldebaran itself—even though Aldebaran is not a genuine 'Hyad', and merely lies between the cluster and ourselves.

There are plenty more of these loose clusters within the range of binoculars. There is, for instance, Præsepe in the rather dim constellation of the Crab; it is nicknamed the Beehive, and is well worth finding, though it is not so rich as the Pleiades. Not far off is another cluster, which has no individual name and is known to astronomers as Messier 67. It is just visible to the naked eye as a dim blur (though I admit that I find it very difficult), but with binoculars it is unmistakable. In the summer we have the so-called 'Wild Duck' cluster in the small group of Scutum, the Shield, which is fan-shaped, and lies in the Milky Way; we have, too, the double cluster in the sword-handle of Perseus, again in the Milky Way.

32

As well as its stars, the Galaxy contains huge clouds of dust and gas which we call nebulæ. Not many of them are easily seen with binoculars (at least in recognizable form), but there is one in Orion, below the three stars of the Belt. With the naked eye it is visible as a faint glow, but binoculars show not only the nebula but also some of the stars contained in it, which make the material shine.

Nebulæ of this kind are of tremendous interest to astronomers, because it is here that fresh stars are being formed. The process is not quick; but there is little doubt that a star begins its career by condensing out of the nebular material— and more than five thousand million years ago, our own Sun was formed in precisely the same way.

Finally, there is much to be learned from using binoculars to seek out some really remote objects. In the rather large, sprawling constellation of Hercules, well on view during summer evenings, we have a star-cluster which is quite different from the Pleiades; it is symmetrical, and it is easy to understand why it is called a 'globular' cluster. It is barely visible with the naked eye, but with binoculars it is quite distinct, even though a telescope is needed to resolve it into stars. It lies near the edge of our Galaxy, and it is so far away that its light, moving at 186,000 miles per second, takes over 22,000 years to reach us.

Yet even this is not the limit of our binocular range. Not far from the Square of Pegasus, in the constellation of Andromeda, may be seen yet another dim blur; with binoculars it is unmistakable, but it takes an effort of the imagination to realize that what we are seeing is a true 'star-city'—an independent galaxy much larger than ours, and containing more than our own total of a hundred thousand million suns. With binoculars (or for that matter, with most telescopes) it is by no means spectacular, and photographs taken with giant instruments are needed to bring out its spiral form, but it is scientifically one of the most important objects in the sky. Its light has taken over two million years to reach us, so that we see it today as it used to be more than two million years ago.

This very brief review may, I hope, serve to show the kinds of wonders that are visible in the night sky with very modest and inexpensive equipment, so let me now give another summary of my personal views. If you have only a limited amount of money to spend, do not buy a telescope unless you happen upon a true bargain; instead, invest in binoculars. For an 'all-purpose' pair, 7 × 50 is admirable; a magnification of up to 12 is satisfactory inasmuch as the binoculars can be steadily hand-held, but with any greater magnification the cost rises sharply and a mounting is needed. An outlay of £15 should be ample for the kinds of binoculars I have been discussing here; and when you have made your purchase, you will find plenty to see.

4 Refracting Telescopes

Now let us turn to telescopes in more detail. Since the refractor is the more familiar form it may be wise to deal with it first, even though reflectors are probably more popular nowadays among serious amateurs—whatever may be the owner's age.

As we have seen, a refractor collects its light by means of an object-glass, and the magnification is carried out by the eyepiece. The distance between the object-glass and the point at which the rays of light are brought to focus is known as the focal length. The effectiveness of the telescope depends on the size and quality of its object-glass, and an aperture of 3 inches is, in my opinion, the minimum for real usefulness, though smaller telescopes will give attractive low-power views.

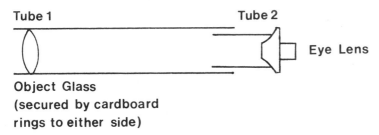

A HOME-MADE TELESCOPE. *The method is described in the text. The main difficulty nowadays is in obtaining suitable lenses—which are not nearly so easy to find as they used to be.*

35

It used to be possible to make a small refractor out of spectacle-lenses and cardboard tubes; and although the resulting instrument would be very imperfect, it would be

A 3-INCH REFRACTOR. *This kind of telescope is particularly suitable for the newcomer to astronomy; the only drawback is the relatively high cost. This photo shows my own 3-inch refractor on a heavy wooden tripod. The pillar of the telescope screws into the metal plate at the top of the tripod. For transport purposes the pillar can be unscrewed from the tripod in a matter of minutes. In fact, the complete telescope-with-tripod can be carried about easily, and I actually keep the instrument indoors. Note the small telescope fixed to the side of the main tube; this is a finder, with a low magnification but a very wide field. Anyone who wants to study the Sun will be wise to buy a refractor rather than a reflector; a 3-inch is an 'all-purpose' instrument, and, in my view, the smallest refractor which is really valuable for astronomical work.*

better than nothing at all. The procedure was to obtain a spectacle lens of about 2 inches in diameter and $1\frac{1}{2}$ feet focal length, and fit it into a tube to serve as the object-glass; the smaller eyepiece lens (say 1 inch aperture, $1\frac{1}{2}$ to 2 inches focal length) was then fitted into a smaller tube which could slide in and out of the first, as shown here. Focusing could be adjusted by sliding the smaller tube, and a reasonably firm mounting could be made out of pieces of wood. Only a few years ago I made a telescope of this sort at a total cost of less than a pound, and it was able to show the lunar craters and the moons of Jupiter quite well. Alas, modern spectacle-lenses are not circular, and are also unsuitable in other ways—though if your budget is very limited it is always worth asking the local optician whether he can find such a lens on one of his dustier shelves!

There can be no doubt that a professionally-made refractor is more convenient than a reflector, because if it is well looked after it needs no adjustment for year after year. Around 1931 I bought a 3-inch refractor for the sum of £7. 50, and it still serves me well; the maintenance required is virtually nil. Unfortunately the modern price-range is depressingly different, and the 1931 cost must be multiplied by at least ten, while—as we have noted—second-hand telescopes are now very rare indeed.

This is a great pity, because a 3-inch refractor is the ideal beginner's telescope. It is powerful enough to be useful, and yet small enough to be portable and 'handy'. It might be thought that a 2-inch would be almost as good, but in fact the extra inch of aperture makes a surprising difference. And with refractors over 3 inches aperture, the cost rises alarmingly; moreover a 4-inch is barely portable, while a 5-inch must be set up permanently, preferably in some sort of observatory.

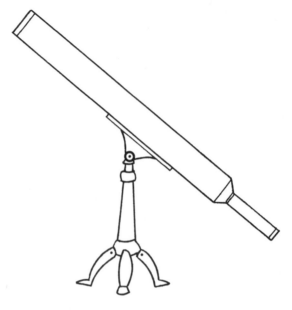

PILLAR AND CLAW MOUNT. *This horrible arrangement is as unsteady as a blancmange. If you buy a telescope on a pillar and claw, I recommend putting the pillar on to a heavy tripod (as shown in the photographs in this book) and consigning the claw to the dustbin.*

The mounting of a telescope is all-important, because if the instrument is unsteady it will be to all intents and purposes useless. My original 3-inch was mounted on what is termed a pillar and claw, which is about as steady as a blancmange. The remedy was to take it off the claw, and put the pillar on to a heavy tripod, as shown on page 36, which was far more satisfactory.

No astronomical telescope, even the smallest, can be used without a mounting. The troubles involved in holding up a heavy pair of binoculars are increased out of all recognition; the telescope must be kept steady—and it must be capable of being moved very slowly and regularly. This applies as much to the home-made spectacle-lens instrument as it does to the professionally-built telescope. Also, a mounting must be really solid; within reasonable limits, the heavier the better.

For a refractor, a tripod will do; this can be bought without too great a financial outlay—and, of course, anyone who is good at carpentry can make it. The 'joint' carrying the telescope itself, on the top of its pillar, is more of a problem, but will almost always be included with the purchased telescope itself. If you want to have something more complicated—say an equatorial mounting, which can be fitted with a clock drive to compensate for the apparent movement of the sky—the cost goes up; but to deal with equatorials here would be beyond my scope, and the best plan is to read some specialized book.

Any astronomical telescope, whatever its size, should have several eyepieces. A low magnification will give a wide field, and will be suitable for looking at objects such as star-clusters and nebulæ; a higher power is to be preferred for the Moon and planets, where the aim is to see as much fine detail as possible; and a still higher magnification is desirable for use on really clear, calm nights. To find the magnification, simply divide the focal length of the object-glass by that of the eyepiece. For instance, my 3-inch refractor has a focal length of 36 inches; thus with an eyepiece of focal length 1 inch, the magnification will be $36 \div 1$, or 36. A half-inch eyepiece will give $36 \div \frac{1}{2}$, or 72; and I also have

39

a quarter-inch eyepiece, giving $36 \div \frac{1}{4}$, or 144. In theory, it is possible to obtain as much magnification as you like simply by using eyepieces of shorter and shorter focal length; but with each increase the image becomes fainter, and if I used, say, an eyepiece of 1/10 inch on my refractor, giving a magnification of 360, the image would be so dim that it would be useless. To obtain a power of 360, you must use a larger telescope, which will collect more light.

One very useful addition to a telescope is a finder. This is simply a very small telescope with a very wide field, mounted on the main instrument. The object to be studied is brought into the field of the finder, which is easy enough; if everything is properly adjusted, the object will also be visible in the eyepiece of the main telescope. A finder is not strictly necessary, but it is certainly convenient, and I would hate to be without one.

The thing *not* to do with a refractor is to tinker about with the object-glass. Most objectives are made up of two or more lenses set together; and if they become out of adjustment, there may be endless trouble. There is, in fact, no need to touch them at all, and on no account rub them if they seem to be dusty or dirty. Blow the offending dust off; if essential, wipe very lightly with silk cloth, but remember that a lens is a delicate thing.

If you find that your refractor is not giving good images, the first thing to suspect is the eyepiece; using a high-quality objective with a poor eyepiece is rather like using a good record-player with a bad needle. Change the eyepiece, and see whether things are better. If not, and the performance is still not up to standard, it is wise to take the telescope to an expert for checking (see page 78). Probably it is only a matter of adjustment, though there is no cause for a refractor to give trouble unless it is mishandled.

I do not propose to say anything here about the more complicated types of mountings, which are very convenient for ordinary observation and are essential if you want to take any photographs through your telescope; these put up the cost at once. Meantime, it is enough to repeat that if you have the chance to buy a good 3-inch refractor at reasonable cost, you will be wise to accept—though before coming to any definite conclusions, please read the next chapter.

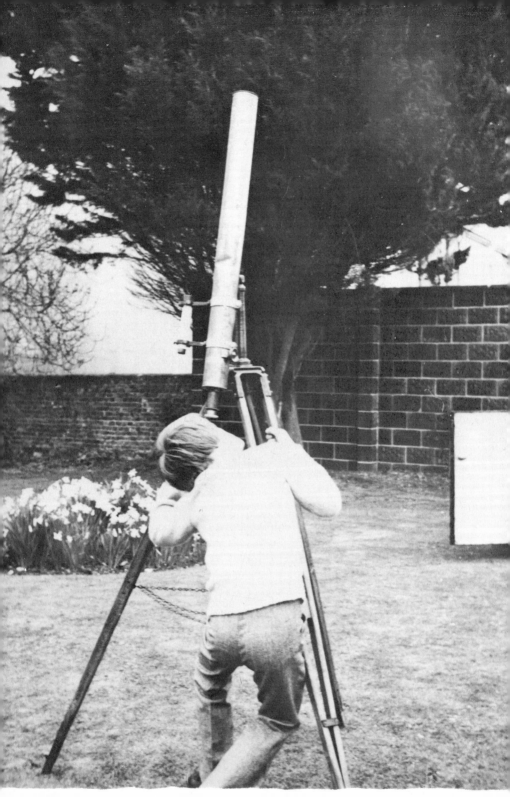

5 Reflecting Telescopes

Reflectors are of various kinds, but I propose to deal here only with the Newtonian. The principle is shown in the diagram on page 19. Obviously, the flat mirror blocks out a little of the light coming down the tube; but the loss is not serious, and there is no practicable way to avoid it.

I have already commented that inch for inch, a mirror is less effective than a lens. A 3-inch refractor is a useful telescope; a 3-inch reflector is, to be honest, fairly useless. A 4½-inch reflector is just about on the borderline, but for most purposes an aperture of 6 inches is the required minimum. This brings one up into the £70 to £90 range, and there may be a straightforward choice between, say, a 6-inch reflector and a 3-inch refractor. The choice depends upon the personal views of the observer—and, most important of all, what branches of astronomy interest him most.

< ONE OF THE DISADVANTAGES OF A SMALL REFRACTOR. *With a refractor, such as the 3-inch shown here, one trouble is that when the telescope is pointed at an object high in the sky the eyepiece is low down. This means that the observer has to crouch down in an uncomfortable position. The trouble can be overcome by putting the telescope on a much taller tripod—but this makes for clumsiness and lack of portability, and when the object to be viewed is near the horizon the telescope will have its eyepiece so high up that the observer has to mountaineer.*

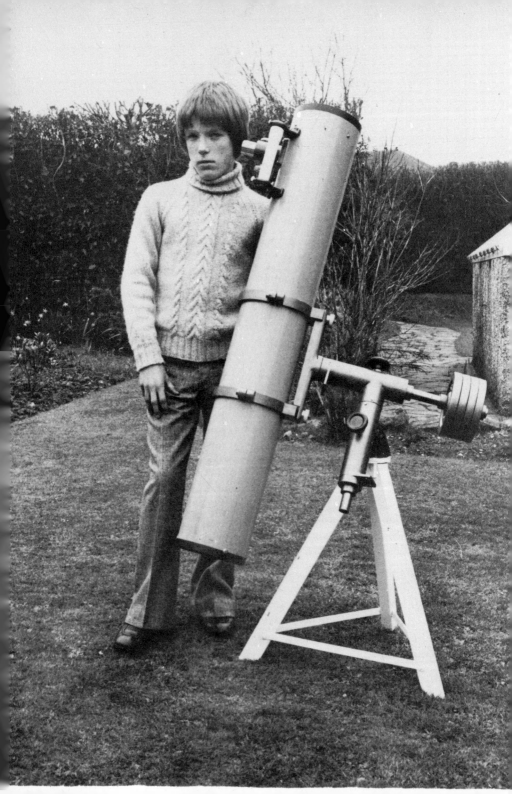

For example, if you want to study the Sun, you need a refractor: if you have a general interest, and want to see rich star fields and overall views of the Moon and planets, either a refractor or a reflector will do—the emphasis being upon 'handiness'; if you are anxious to see the Moon and planets in detail, you need as much magnification and hence as large a telescope as you can get.

A 6-inch reflector has a definite advantage over a 3-inch refractor, and in some ways it is easier to use. When a star is high up, the refractor-owner has to become an expert muscle-twister, because the tube is pointing skyward and the observer has to put his eye to the lower end—which often means bending one's body into a hoop. In this respect reflectors are much more comfortable. On the other hand,

< THE 6-INCH REFLECTOR. *This solid-tube 6-inch reflector is an ideal instrument for the beginner (unless he wants to study the Sun). It can be carried about within certain limits, though of course it is not truly portable in the same way as a 3-inch refractor; for one thing it is set up on an equatorial mounting.*

The tube is of fibreglass (though in fact the tube may be of virtually any material: metal, wood or even hardboard—and some reflectors are tubeless). The tube can slip round in the rings which hold it, so that the eyepiece can be put in the most comfortable position for the observer. The finder, at the top of the main tube, is again of low magnification but with a wide field; with a reflector, a finder is really a necessity, since one is looking 'into' the main tube rather than sighting the object along it, as with a refractor.

The tripod in this instance is made of wood, and is firm; the mounting is of the German type, and the weight to the right-hand side is merely a counterweight to balance the telescope.

they have the great disadvantage that they are apt to go out of adjustment, and they are frequently 'temperamental' for no apparent reason. Moreover, the mirrors have to be coated with a very thin reflecting layer of silver or aluminium, and this lasts for only a limited time before it tarnishes and has to be renewed. I live within a few hundred yards of the sea, and I find that I have to have the mirrors of my reflectors re-aluminized at least once a year, which is annoying even though it costs a very few pounds. Of course, people who live inland, where the air is less salty, will find that their mirror coatings last longer, but sooner or later the re-aluminizing or re-silvering process has to be carried out.

Unless the telescope can be kept permanently set up, and given some protection against the weather, it must be carried outdoors when it is to be used, and taken in again afterwards. Here the reflector is at a disadvantage, because

TELESCOPE MOUNTINGS—THE ALTAZIMUTH. *The main requirement for a telescope mount is that it should be firm. If you decide to make one, a good rule is to work out the maximum weight which you think will be necessary for the mounting—and then multiply by three!*

The simplest form of mounting is the altazimuth, shown in the next two photographs. The 3-inch refractor is mounted on a pillar which screws into a heavy tripod, and can be moved freely in any direction, either up and down (altitude) or east and west (azimuth): hence the name altazimuth. The disadvantage is that the telescope cannot be driven to compensate for the Earth's rotation, and there are two movements to be taken into account all the time, so that constant adjustment has to be made—particularly when using a high magnification.

the mirrors may be jolted out of their proper position, and the slightest misalignment means that the image will be blurred. Everything has to be 'just right'. This is not the place to explain how the adjustments are made, but anyone who buys a reflector should take the trouble to learn how to do them.

The effectiveness of the telescope depends entirely upon the two mirrors (plus the eyepiece, of course), and of these the more expensive is the main mirror at the bottom end of the tube. If you look at it casually it will seem flat; but really it is curved, as otherwise it would not bring the light-rays to a focus. The shape of the curve has to be remarkably accurate, and any serious fault means re-figuring the whole mirror, which is a task for the expert and is expensive if done professionally. On the other hand, it is never wise to condemn a reflector without checking it very carefully. In most cases, poor performance will be due to faulty adjustment of the optics.

With a reflector, a finder is really necessary, and for an instrument of 6 inches aperture or larger one needs 'slow motions' as well. Pushing the telescope to follow an object

The 12½-inch reflector is also on an altazimuth stand. The > *solid metal tube rests in a cradle, and there are slow motions; the telescope can be moved in azimuth by twisting the handle which hangs down from the main plate, and in altitude by adjusting the thread of the rod which runs from the top of the tube through to the projecting rod below. Not many telescopes of this size are on altazimuth mountings, but they do have some advantages, inasmuch as they are uncomplicated. The heavy tripod is bolted down to the concrete base upon which David is standing.*

creeping across the sky is not easy, and it is far better to have some arrangement by means of which the telescope can be moved, slowly and steadily, by turning a rod or a handle.

Mountings are of various types. Some of the common patterns are shown on pages 46-53. The German type balances the telescope by means of a counterweight; the Fork is, in my view, rather better, though it has to be very solidly made; the English is effective but clumsy, and cannot be shifted around.

What of the telescope itself? The tube may be made of any material—and it need not be cylindrical; some are square, and my own 15-inch reflector has an octagonal 'tube' made of wood. There need not be any tube at all, and many reflectors are 'skeletons', as shown on page 54. This makes for lightness—an important factor if one has to carry the telescope around—but the framework of the skeleton must be really firm, as otherwise the mirrors will go out of adjustment at the slightest touch. In some patterns the flat mirror at the upper end of the tube is merely clipped

< TELESCOPE MOUNTINGS—THE GERMAN. *The mounting shown here has a counterweight, and is called the German. The 6-inch reflector is fixed to one side of the 'polar axis' upon which David has his hand, and the weight is balanced by a counterweight on the opposite end of the polar axis. The polar axis points to the pole of the sky, and so when the telescope is moved in azimuth (east-west) the 'up and down' movement of the object under observation looks after itself. The knob on which David has his hand is a clamp to fix the telescope in altitude once the object has been brought into the field of view. Of course, care must be taken to see that the axis really is pointing toward the pole!*

51

on to part of the skeleton by a sort of strut, though this is definitely not something which I would recommend.

Any Newtonian reflector of aperture over 6 inches must be regarded as strictly non-portable, and the cost of the telescope rockets up with every extra half-inch. Mirrors cannot be mass-produced if they are to be of good quality. There is no short cut, and they have to be finished off, at least, by hand. Yet mirror-making is not so difficult as it might sound, and many amateurs have tackled it with success. I know of one fifteen-year-old enthusiast who has constructed an excellent $8\frac{1}{2}$-inch reflector, and uses it on every clear night.

To make a mirror, you have to buy two glass disks or blanks, and then grind one against the other until the shape of the curve is perfect. There are various optical tests which can be carried out—and which are amazingly sensitive. Yet the whole process takes many tens of hours, and the beginner has to be prepared for several disappointments; there will be times when he is tempted to wrench the mirror off the bench and stamp on it. Moreover, grinding a mirror is not something which should be started when G.C.E. examinations are looming ahead. Alternatively, one can buy the optics

TELESCOPE MOUNTINGS — THE FORK. *This is the $15\frac{1}{2}$-inch* > *reflector in my observatory at Selsey, and is contained inside a dome, with a slit which can be rolled back. The tube, only partially enclosed, is of wood; as can be seen from the photograph, the telescope can be swung through the 'prongs' of the fork. This is, of course, an equatorial, and an electric motor drives the telescope around so as to keep pace with the sky. At the top of the tube can be seen the flat mirror, supported by four thin but very firm metal arms.*

A SKELETON REFLECTOR. *A tube is not necessary for a reflector, since the only real need is to keep all the optics in the correct position. There are some telescopes which are entirely tubeless, such as the 12-inch reflector made and used by Commander Hatfield in Sevenoaks. The 6-inch reflector shown below is used by F. R. Spry, Selsey amateur astronomer. It is on an altazimuth mounting, and when the tube and the tripod are separated it is easily portable. The framework is of wood, and here too there is provision for rotating it in its cradle so as to make sure that the eyepiece can be put into a comfortable position. The photograph on the right shows the top of the tube, with the mounting for the finder, which is held in rings attached to a framework screwed into the wooden part of the telescope. There is much to be said for this sort of design; there can be no fear of tube-currents (that is to say, air swirling about inside the telescope and upsetting the definition). If there is any trouble from stray light, the skeleton can be surrounded with cloth or something of the kind.*

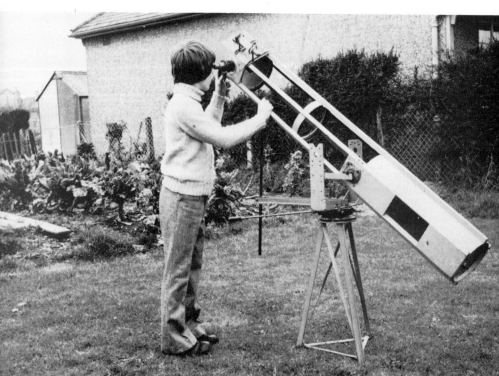

and make the mounting; this simplifies things, though naturally it does put up the cost.

Anyone who wants the most powerful telescope he can obtain for as low a cost as can be managed would, I feel, do well to consider a 6-inch reflector. Make it by all means, if you have the practical ability and enough spare time; if you do not feel equal to tackling the optics, buy the mirrors and mount them; if (like myself!) you are clumsy with your hands, buy a ready-made telescope. This is, of course, assuming that you can afford the outlay of something not far short of a hundred pounds. The comforting thought is that the expense is non-recurring. If you look after your telescope, it will serve you all through your lifetime—and if you do decide that you want to change to something larger, you will certainly have no difficulty in making a sale.

6 Housing the Telescope

Telescopes tend to be rather awkwardly-shaped things, and it is not always easy to decide where to keep them. There is no real trouble with a refractor, which can generally be taken off its mount and stored in a box if need be. The reflector is more delicate, and care must be taken not to jolt the mirrors out of alignment when the telescope is being moved around.

A 3-inch refractor is portable; so is a 6-inch reflector of the Newtonian type—always provided, of course, that the mountings are simple, and do not involve equatorials and clock drives (which I do not propose to discuss here). Anything larger does need a permanent site; and for that matter it is convenient to have even a 6-inch in its own observatory if possible.

The term 'observatory' may sound rather frightening, but in fact a telescope housing can be easy to make and use. One suitable pattern is the run-off shed, as in the photograph here. The shed is in two parts, mounted on rails; when the telescope is to be used the parts of the shed are pushed back, leaving the instrument in the open. Angle-iron can be used for railing, and the whole structure can well be of hardboard. I have used a run-off shed for my 12½-inch reflector ever since 1947, and it has given no trouble at all. I am less fond of a run-off shed with a door at one end, because the door must be either hinged or removable, and in either case it is a nuisance.

A RUN-OFF SHED. *This is the run-off shed for the $12\frac{1}{2}$-inch altazimuth reflector at my Selsey observatory. The shed is in two parts, which meet in the middle. Each side runs back on rails. The shed has now been in operation for over 25 years, and has never given the slightest trouble. It is of wood, but other run-off sheds may be of hardboard, plastic, or indeed almost any material which is weatherproof. Angle-iron will serve quite well as a rail. My own rail is embedded in concrete, and this is probably the best method, though there is no objection to having the run-off shed on ordinary wheels and simply rolling it back.*

The main advantage of a run-off shed is that it is cheap and easy to make. I recommend the design shown here rather than a shed with a door at one end; if the door is hinged it is liable to flap, and if it is removable it can be awkward to replace in the middle of the night when the user's hands are cold.

57

Another kind of observatory involves a building with a roof which runs off on rails (see photograph below). This is excellent for a refractor, but takes up a good deal of room, and looks rather clumsy. Then, of course, there is the proper

A RUN-OFF ROOF OBSERVATORY. *This observatory houses my 5-inch refractor at Selsey. The framework is of wood, and the sides are plastic. The roof runs back on rails, on to a support which is shown to the left-hand side of the photograph; the roof is in the run-back position. This arrangement is excellent for a refractor, but not for a reflector, where the high sides would cut off much of the sky. (The solution would be to put the rail lower down, so that the whole top part of the observatory would roll back instead of merely the roof.) The other disadvantage is that it takes up a good deal of room, and looks rather clumsy, but for my 5-inch refractor I have found it ideal.*

AN OBSERVATORY DOME. *Here, the dome itself is made of wood, and so is the roof. The upper section (everything above the octagonal wooden wall) is rotatable; this involves a perfectly circular rail. The slit in the roof can be opened, and so can the window, as shown here. The instrument inside the dome is an $8\frac{1}{2}$-inch reflector on a German-type equatorial mounting.*

The glass windows were originally put in to make the observatory look decorative. In fact, few observatories have glass of this kind, which is reputed to heat up the interior—though I have not found any trouble on this score. The whole building stands on a concrete base, and I have found no need to fasten it down. The structure was made in parts, so that it can be dismantled, transported and reassembled without undue difficulty, though great care must be taken.

59

A FULLY ROTATING OBSERVATORY.   *This observatory was made and is used by F. R. Spry, at Selsey; inside it is an 8½-inch equatorial reflector on a fork mounting. The building is square; the roof is easy to open, and the user can walk straight inside without having to crouch down below a rail. The rail is at ground level, and the entire building can be rotated. The observatory is highly efficient— but anyone who sets out to make a fully-rotating observatory must take extra care, as there is more mass to be moved than in a rotating-top arrangement. The only hazard I can find is that since the door moves round, one never knows where it will be at the end of an observing session after the slit has been closed, and it is only too easy to step out into, say, a cabbage patch, such as the one shown on the left.*

In general, I would say that a rotating observatory of this kind, if as well made as the building shown here, is excellent for any reflector of 8 inches aperture or below.

dome, shown here; the dome itself is rotatable, and there is a slit which therefore moves round so that the telescope can see through it. Building a dome is purely a problem of carpentry and mechanics. The worst difficulty is, usually, in making or buying the circular rail on which the dome revolves; much depends on the availability of a friendly blacksmith. An alternative pattern is to make the entire building revolve, though unless everything is very carefully made and kept oiled there is a tendency to stick.

Of course, a dome is bound to be costly even if it is almost or wholly home-made, but a run-off shed is pleasingly cheap. On the other hand there are many people who have no space to keep a telescope permanently set up or to build an observatory of any kind. It is often convenient, and possible, to make a concrete or iron pillar upon which the telescope, on its mounting, can be fixed; the drawback here is the need for dismantling the telescope every time observing is finished. Covering a telescope with a tarpaulin or anything of the sort is not a good idea, since it is the reverse of damp-proof. Actually, damp will not do a refractor a great deal of harm, but the mirror of a reflector will very soon tarnish. All in all, I would definitely not recommend leaving a telescope outdoors unless it is in some kind of observatory.

I may add that I have had many letters from people who feel that they can observe from indoors, through a window, or alternatively set up an observatory on the roof of the house in which they live. Alas, it is hopeless to try to use a telescope from a window; the heated air from the room is bound to swirl up outside, and the image will be hopelessly unsteady. (Quite apart from this, it is only a question of time before an eyepiece or even the whole telescope is dropped through the window!) Roof-top observatories are useless for the same

61

reason, plus the fact that they are open to the full fury of any wind which may be blowing. If you want to set up an observatory, or a pillar for your telescope, keep it as far as you can away from any building which is warmed. There is, unfortunately, no choice but to put up with awkward things such as tall trees and street-lamps.

To sum up:

If you have a 3-inch refractor or a 6-inch reflector, or any smaller telescope, there is no need for an observatory, because the instrument is portable enough to be kept indoors and taken out when needed. If you have a larger telescope (or, of course, if you have the means and the wish to make an observatory for the telescope you have), then consider a run-off shed; it is relatively inexpensive, and it works well. With more money to spend, and a suitable site, a dome is of course ideal. But do not leave your telescope outside, unprotected, at the mercy of the weather—and if you have to carry it around, take the greatest care!

7 Using your Telescope

Anyone equipped with an adequate telescope—by which I mean a refractor of at least 3 inches aperture, or a 6-inch reflector—will find that there is endless interest in the sky. The first step is to look around and become really familiar with the telescope; then, if you feel so inclined, select one or more lines of observation which interest you particularly.

All I propose to do here is to give some hints for the absolute beginner at telescopic work—and to make a few suggestions. Let us assume, then, that you have your 3-inch refractor or 6-inch reflector, or something larger.

THE SUN

As I have said more times than I can count, there is one golden rule about looking at the Sun: *don't.* Direct observation, with any telescope or even binoculars, is horribly dangerous, and I am no believer in the dark caps which, as some manufacturers claim, can be fitted over the telescope eyepiece to cut down the heat and light to an acceptable level. Such caps are never really safe, and are apt to splinter without warning.

Therefore, use the projection method, as shown here. Point the telescope toward the Sun, without looking through it—and preferably with a tin or cardboard cap over the tube. Then remove the cap, and hold a screen behind the eyepiece. The Sun's image will be very clearly seen, together

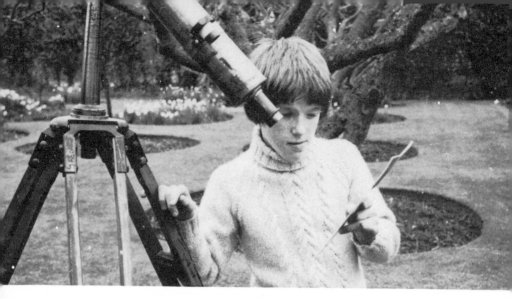

PROJECTING THE SUN. *Anyone who wants to study the Sun should have a refractor rather than a reflector. Dark sun-caps, placed over the eyepiece and used for direct viewing, are emphatically not to be recommended, since they can never give full protection, and are apt to splinter without warning. The golden rule is: never look direct at the Sun through any telescope, or even a pair of binoculars.*

The procedure is being demonstrated by David in the photograph. Point the telescope sunward, without looking through it; then hold a white screen behind the eyepiece—the Sun's image will be clearly shown, together with any spots which may be on view. The serious observer will fix up a 'shade' on the end of the main telescope tube, to cast a shadow on to the screen and reduce the glare caused by scattered light, and will also fix up a proper projection box which can be attached to the telescope and replace the simple method of hand-holding. However, the same principles apply; projection in some form is the only safe and sensible method of studying the Sun with a small telescope. Incidentally, larger telescopes collect more light than is needed—with a consequent increase in heat. If I ever use my 5-inch refractor for the Sun, I put a 'stop' over the object-glass and reduce the aperture to 4 inches or even 3; and to use my large reflectors for looking at the Sun's image would be stupid in the extreme.

with any of the dark patches called sunspots which happen to be on view. Note, by the way, that the illustration shows a refractor being used. A reflector is not suitable for solar work—and so if your main interest is the Sun, obtain a refractor if you possibly can. Even a 2-inch will show the sunspots quite well.

The Sun spins on its axis once in between three and four weeks, so that the spots can be seen to shift in position from one day to the next as they are carried around. Since the Sun's surface is gaseous, no spot can last for very long; but it often happens that a spot-group can survive to make several crossings of the disk. Plotting the positions and forms of the spots is always interesting.

THE MOON

Here, the main interest with a small telescope is in identifying the various features—the waterless seas, the mountains, the valleys, the craters and so on. My own method was to take an outline map of the Moon and then, using my telescope, make one or more drawings of each named formation. The drawings themselves were useless—except to me; by the time I had finished (and the whole project took me over a year) I really knew my way around the Moon. It is worth making at least two sketches of every major crater, on different nights, because the differing angle of sunlight makes a tremendous difference to what is seen. For instance, the great 90-mile crater Ptolemæus, near the centre of the disk, is striking when the Sun is rising or setting across it, because its floor is largely in shadow; but near full moon it is not easy to locate at all, and other craters become even harder to find.

Sometimes the Moon passes in front of a star, and hides or 'occults' it. Because there is no atmosphere round the Moon's edge, the star seems to snap out instantaneously. Occultations are always worth watching—and timing, if you have a stop-watch.

THE PLANETS

Here, the scope for the owner of a small telescope is somewhat restricted, even though a 6-inch reflector is big enough for useful work to be done at least with regard to Venus and Jupiter.

Of the two inner planets, Mercury is hard to pick out; you have to catch it just after sunset or just before sunrise, when it is best placed—and if you do locate it, you should be able to see the phase (crescent, half or three-quarter). Venus is glaringly obvious, and here too the phase can be seen; in fact it cannot be overlooked, even with a low magnification, though nothing much else will be visible. Mars is a planet which needs high magnification if it is to be well observed, and there is also the point that it is favourably placed for only a few months in every alternate year; but your telescope will show the main dark features (not now thought to be vegetation-tracts!) and also, when suitably placed, the white polar cap over whichever pole happens to be facing the Earth. However, Mars is never an easy object to study, and do not be disappointed if you see almost nothing except for a small red disk.

Jupiter is more rewarding. Here we have not only the four bright moons or satellites (Io, Europa, Ganymede and Callisto) but also the cloud belts crossing the planet itself; and your telescope will show the celebrated Great Red Spot

66

whenever it lies well on the Earth-turned hemisphere. Jupiter spins round in less than ten hours, so that the features change noticeably over short periods. Make disk-drawings at intervals of half an hour or so, and plot the changing positions of the satellites.

Saturn, with its ring system, is certainly the loveliest object in the whole sky. When the rings are 'wide open', as in the early and mid 1970s, your telescope will show them well; you will even see the dark gap known as the Cassini Division, which separates the two bright rings. Note also the shadows cast on to the rings by the disk, and on to the disk by the rings. The belts on Saturn are less obvious than those of Jupiter, but you should be able to see at least one; and there are also several satellites. Titan, the largest of Saturn's moons (and, incidentally, the only satellite in the Solar System known to have a reasonably dense atmosphere) is very clear with a 3-inch refractor; and Iapetus, Rhea and sometimes Tethys and Dione are also visible, though admittedly the two latter are not easy with an aperture below 6 inches.

It is also worth noting that when Saturn's rings are turned edge-on to us they cannot be seen with small telescopes; even in large instruments they show up as nothing more than a thin line of light. The rings will be edge-on again around 1980, and for a few years Saturn will lose the major part of its beauty.

Uranus, the seventh planet, can be located easily with the help of a star-map; indeed it is dimly visible with the naked eye. Your telescope will show that it shows a small, rather greenish disk. The outermost giant, Neptune, is also within binocular range, though it looks very like a star. A much larger telescope is needed to show the last planet, Pluto.

67

COMETS

A comet is not a solid body; it is made up of very small particles together with thin gas. Most comets move round the Sun in very elliptical paths, and we can see them only when they are moving in the inner part of the Solar System. The comets with short periods are rather faint, though a small telescope will show some of them when best placed; they look like dim patches of light. Larger comets, with tails, have periods so long that we never know when to expect them—but make the most of any bright comet which comes into view. Unfortunately comets are unreliable things; many people will remember the case of Kohoutek's Comet of 1973–4, which was expected to become brilliant, but failed to do so, even though it was visible with the naked eye when at its best and stayed within binocular range for some time.

There is one bright comet, Halley's, which comes back every 76 years; but it is not due again until 1986.

STARS

Here again we have plenty of variety. Your telescope will show a great many double stars; using a star-map, locate the various pairs, and note their separations and their colours. Perhaps the loveliest of all is Albireo, in the constellation of Cygnus (the Swan), which is made up of a yellow star attended by a bluish-green companion. Mizar, in the Great Bear, is easily split; with a low power the Mizar pair is in the same field as Alcor, mentioned earlier, and there is a third star between Alcor and Mizar itself. (This third star is more remote; it lies 'in the background', and has no real association with Mizar or Alcor.)

Variable stars, which brighten and fade either regularly

or irregularly, make up an important and fascinating branch of amateur astronomy. The procedure is to identify the variable, and then compare it with nearby stars which do not alter; the brightness of the variable at the time of observation can then be worked out. This type of work makes up an excellent and very useful project, but to go into details here would be beyond our scope.

CLUSTERS AND NEBULAE

There are star-clusters in plenty within range of your telescope. Some, such as the Pleiades, are so large that they are probably best seen with binoculars; but the smaller, less brilliant clusters make splendid telescopic objects, and a good star-map will show where to find them. There are a few globular clusters, that in Hercules being the best available from Britain; though binoculars show it only as a hazy patch, the telescope reveals that it is made up of stars. The open clusters are probably more spectacular; identify them, plot their positions, and—if you have enough patience— sketch their main stars.

There are also a fair number of gaseous nebulæ, though none can rival the Great Nebula in Orion's Sword. Here, sketching is well worth while. Another favourite object is the so-called Ring Nebula, near the brilliant blue star Vega (page 30) which lies directly between two naked-eye stars, and so is easy to locate; in a small telescope it is dim, but greater power shows it to be rather like a faint, luminous cycle-tyre. To be accurate, it is not truly a nebula, but simply a faint star surrounded by a huge shell of thin gas. A few more of these rather misleadingly-named 'planetary nebulæ' will be within your range.

69

Lastly, there are the galaxies—those great independent star-systems, of the same basic type as the Milky Way galaxy which contains our Sun and its family of planets. The Andromeda Galaxy, already described, is the most famous of them, but there are a few more which are bright enough to be seen with a small telescope—even though I have to admit that they are disappointing, and no details can be seen in them.

All this has, I hope, given you some ideas—and this leads me on to my final topic: the School Project.

8 School Projects

Many schools of today place great emphasis on Projects, in which the candidate selects a subject (or has it allotted to him) and then has to fend for himself. Astronomy is something of a favourite, and I have long since lost track of the number of letters I have had about projects of all kinds. I do my best to answer them all, though it means spending many hours at my typewriter!

The most difficult inquiries to cope with are the vague ones. 'I am doing a school project on astronomy, so will you please send me some information, together with pictures and posters?' This is a typical letter (others in the same vein omit the word 'please'), but of course a full answer would mean writing a book. Moreover, the whole idea of a school project is that the candidate should do the work himself, without asking to be spoon-fed. The best answer is to suggest some books which will provide a background knowledge of what astronomy is all about. As I have pointed out earlier, the local public library will be able to provide them, and there is no short cut. (I may add that I give just the same answer to teachers, both trainee and qualified, who write to me in similar vein.) If there are any special queries, then there will generally be someone at hand to help.

In a general project, the background information should be obtained first; then, if the idea is to deal with—say—the Solar System, an account of it can be written up in an exercise book, and any available illustrations can be used. For an ambitious project photographs can be bought, though admittedly this costs money; line drawings can be attempted by the candidate himself to illustrate various points, such as, say, the phases of the Moon or oppositions of Mars. A project book of this kind can be made very attractive.

There are, too, people who want to do some original work for their project—and this is certainly a good idea. There are many possibilities, depending upon what equipment (if any) you have available to use. Without optical aid, you can produce constellation diagrams and descriptions; not so long ago I marked one project which dealt with the colours of bright stars as seen with the naked eye—and it was very good indeed, because the author had been into the whole matter quite deeply and had carried out tests. It was, for instance, interesting to find that whereas most people described Aldebaran as orange, some observers called it red, and others saw it as nothing more than deep yellow, with only a hint of orange. Meteor-watching is another branch open to the naked-eye observer, though admittedly it means spending hours outdoors on nights which may well be cold.

No telescope is needed for meteor-recording; indeed, a telescope would be useless! The procedure is to keep a careful watch on the sky at suitable times and in suitable directions, and note any meteors seen: list the time, the duration of visibility, maximum magnitude (compare with suitable stars whose magnitudes you know), colour, and any unusual characteristic. There are plenty of meteor showers each year, of which the most reliable is that of the Perseids (late July to mid-August), and plotting the tracks on a star-chart will show the position of the 'radiant'—that is to say, the position in the sky from which the meteor seems to come. A list of meteor showers can be consulted (see p. 77), though there are many 'non-shower' meteors as well. Meteor-recording tends to be laborious, but it makes a very good school project.

With binoculars, the scope is increased again; double stars can be identified and described, and clusters and nebulæ treated in a similar way. With variable stars, some genuinely useful research can be carried out. The same applies to the telescope-user, and here, of course, the Moon and planets can be studied as well. There seems no point in going into further detail, because I hope that what I have said in the previous pages will give you as many ideas as you need.

If you have a camera, the project book can be made to look more attractive by the inclusion of suitable photographs —of telescopes, binoculars, or observatories, for instance. Provided that a time-exposure can be made, pointing the camera upward during a dark, clear night and leaving the shutter open for ten minutes or so will produce excellent pictures of star-trails. And if you want to photograph the Sun, merely project the solar disk on to your screen as I have described on page 64, and then photograph the screen. Any sunspots on the disk would be recorded without difficulty.

Certainly a school project on astronomy is well worth doing, and it offers a tremendous range of opportunities— but, as always, make sure to put your ideas in order first; and the only real way to do this is by reading books. Only then can you hope to produce a worth-while project. If you decide to attempt it, I wish you the very best of success.

Appendices

1. A NOTE ON METRICATION

In the text, I have given telescope apertures in inches, which is the measure usually used by suppliers, but metric equivalents may be useful. Those used in this book are as follows:

$2\frac{1}{2}$ inches $= 63$/mm
3 inches $= 76$/mm
4 inches $= 102$/mm
5 inches $= 127$/mm
6 inches $= 152$/mm
$8\frac{1}{2}$ inches $= 216$/mm
$12\frac{1}{2}$ inches $= 317$/mm
$15\frac{1}{2}$ inches $= 394$/mm
36 inches $= 914$/mm

2. THE G.C.E. 'O' LEVEL

For some years now astronomy has been a subject for the G.C.E. 'O' Level examination. It is not difficult; all that is needed is patience and determination, and anyone who is really keen can learn all that is needed by studying books. Few schools teach astronomy as a class subject, but the examination can always be taken as an external candidate.

One paper is set. What is essential, of course, is a good general knowledge of astronomy; the candidate has to keep an observing log and to carry out several practical projects, but it is not necessary to have a telescope—and only an ordinary knowledge of mathematics is required.

My usual advice, when asked about the examination is to 'go it alone', using books, and asking for help from any available expert when need be. I know of many people who have passed the examination without having any formal instruction at all. The only thing to bear in mind is that to spend too much time on it, to the disadvantage of the really essential G.C.E. subjects, is a mistake.

3. ASTRONOMY AS A CAREER

Anyone can become an amateur astronomer—as I am myself. If you want to make a career out of it, however, remember that all modern astronomy is essentially mathematical; and unless you have mathematical ability, I advise staying as an amateur.

The vital point is that any professional astronomer must have a science degree: that is to say, a B.Sc. (Bachelor of Science) or equivalent. Without such a qualification, no positions in science will be open to you, and it is a waste of time to try.

Oddly enough, the basic degree need not be in pure astronomy. I always say to an inquirer: 'If you are 100 per cent certain that you want to be an astronomer, and you are really good at mathematics, then take an astronomy degree by all means; several universities are open to you, notably London. But if you are only 99·9 per cent certain that your career is to be in astronomy, then start off by taking your first degree in physics—or even pure mathematics.' The reason for this is that a physics degree will serve for most openings in astronomy, while for the graduate in

75

actual astronomy it is 'astronomy or nothing'. Bear in mind, too, that anyone with a physics degree can tack on an extra qualification in astronomy when the need arises.

In either case, the first step is to pass G.C.E. 'A' level in at least mathematics, physics and another science subject, such as chemistry; you also need the usual string of 'O' levels. Only when these passes have been secured can you begin a degree course. And I repeat that for a professional astronomer, a degree is absolutely essential. There is no short cut.

4. ASTRONOMICAL SOCIETIES

If you become really interested in astronomy, I strongly recommend joining a society. The national amateur organization is the British Astronomical Association (Burlington House, Piccadilly, London W.1), which publishes a regular Journal and meets on the last Wednesday of each month from October to June. There is no actual age limit (I joined at the age of eleven) but, of course, some of the papers and programmes are advanced, and it may be as well for the very young enthusiast to start off by joining a local society. There are plenty of these, all of which are listed in the annual *Yearbook of Astronomy* published by Sidgwick & Jackson. If you decide to join, you will make plenty of friends —and you may well have the chance of using a telescope if you do not have one of your own.

5. SOME BOOKS

Many hundreds of books on astronomy have been published in recent years, and the following list is not intended to be

APPENDICES

anything more than a very general comment upon some of the books which I have found very useful.

Norton's Star Atlas (Gall & Inglis) is absolutely invaluable. It gives complete charts of the whole sky, together with a lengthy text. Admittedly it is designed for the serious observer; but nobody should be without it. There are many simpler star-maps, and also the useful revolving planispheres (Philip & Son) which can show the view of the sky for any time and date of the year. There are two maps of the Moon designed for crater-identification, one by T. G. Elger (Philip & Son) and the other, a wall map to a scale of two feet to the Moon's diameter, by myself.

Astronomy for Beginners, by Henry Brinton, is an excellent general outline. Rather more elementary is *Simple Astronomy*, by Iain Nicolson (Nelson); and *The Solar System*, by E. A. Beet (Blackie) will be found most useful. Lists, tables and general notes are given in *Astronomy Data Book*, by J. Hedley Robinson (David & Charles) and the much more extensive *Encyclopedia of Astronomy*, by G. E. Satterthwaite (Hamlyn). For the more advanced reader, it is hard to better the classic *Astronomy* by Baker and Fredrik (Van Nostrand). Perhaps I may be allowed to add two books of my own: *The Amateur Astronomer* (Lutterworth Press) and *Astronomy for 'O' Level* (Duckworth).

There are many books dealing with telescope-making. *Handbook of Telescope Making*, by N. Howard (Faber & Faber) is detailed and comprehensive; *Telescopes and Observatories for Amateur Astronomers* (David & Charles) is a general survey in which the mirror-making chapters are by T. W. Rackham.

TELESCOPE SUPPLIERS

Nowadays there are many firms of telescope-suppliers, who can provide instruments of all sizes from the very small to the very large. The equipment in my own observatory was set up mainly by two London firms:

Henry Wildey (14 Savernake Road, Hampstead, London NW3) and Fullerscopes (Telescope House, 63 Farringdon Road, London EC1).

All firms can also provide essentials such as eyepieces— and they can undertake repairs and the aluminizing of mirrors.

Index

Index